Ronilson Freitas de Souza
Abraão J B Muribeca
Paulo W P Gomes

Bee honeys produced in Marajó-PA

Ronilson Freitas de Souza
Abraão J B Muribeca
Paulo W P Gomes

Bee honeys produced in Marajó-PA

Phenolic compounds and antioxidant activity

ScienciaScripts

Imprint
Any brand names and product names mentioned in this book are subject to trademark, brand or patent protection and are trademarks or registered trademarks of their respective holders. The use of brand names, product names, common names, trade names, product descriptions etc. even without a particular marking in this work is in no way to be construed to mean that such names may be regarded as unrestricted in respect of trademark and brand protection legislation and could thus be used by anyone.

Cover image: www.ingimage.com

This book is a translation from the original published under ISBN 978-3-330-76025-7.

Publisher:
Sciencia Scripts
is a trademark of
Dodo Books Indian Ocean Ltd. and OmniScriptum S.R.L publishing group

120 High Road, East Finchley, London, N2 9ED, United Kingdom
Str. Armeneasca 28/1, office 1, Chisinau MD-2012, Republic of Moldova, Europe
Printed at: see last page
ISBN: 978-620-8-31574-0

Summary

Presentation

The authors already have extensive knowledge of the quality of honeys produced in the Marajó region. While in their first book "Bee honeys produced on the island of Marajó-PA: Physico-chemical and microbiological quality", they dealt with the physico-chemical and microbiological aspects of honeys produced in this region, in this new work the authors focus on discriminating the content of phenolic compounds and antioxidant capacity of honeys produced in the Marajoara region. The book includes a literature review focusing on the fact that honey is primarily a plant product, hence the importance of correlating its pharmacological activity with its plant origin. The choice to quantify phenolic compounds was due to the great biological activity that these secondary metabolites have shown according to reports in the most diverse literature. It is hoped that this work will legitimize popular knowledge about the use of honey as a component of many home remedies. It is also hoped that local beekeeping and meliponiculture techniques will be able to better exploit the region's wealth of raw materials so that, combined with the ingenuity of the bees, they can produce honeys with greater therapeutic potential.

Thanks

To **GOD**, for showing us the way to make this work possible.

The University of the State of Pará, for the actions that promote the scientific and social development of Pará.

The Campus XIX of the State University of Parâ and the Natural Products Engineering Laboratory (LEPRON) of the Federal University of Parâ for making their laboratories available for the development of this research.

To the beekeepers and meliponiculturists on the island of Marajó, who contributed samples of bee honeys.

To PIBIC-UEPA for the research support granted during the development of the project.

To all the people who directly or indirectly collaborated with the authors to complete the chapters of this work.

The authors

RONILSON FREITAS DE SOUZA

PhD in Organic Chemistry (UFPA). Professor in the Department of Natural Sciences at the State University of Parà (UEPA) / Campus XIX/Salvaterra-island of Marajó-PA. Research Group on Chemistry, Chemistry Teaching and the Environment (http://dgp.cnpq.br/dgp/espelhorh/0747461930362318) and CV: http://lattes.cnpq.br/0747461930362318

E-mail: (ronilson@uepa.br)

ABRAAO DE JESUS BARBOSA MURIBECA

Chemist. Master's degree in Organic Chemistry (UFPA). Researcher at the Federal University of Parà (abraao_muribeca@hotmail.com);

CV: http://lattes.cnpq.brZ1006873022498893

PAULO WENDER PORTAL GOMES

Chemist. Master's student in Organic Chemistry (UFPA). Researcher at the Federal University of Parà (wenderufpa@hotmail.com);

CV: http://lattes.cnpq.br/6102293278640224

Chapter 1

Honey and its medicinal properties

Honey is a viscous, aromatic product with a generally sweet taste, which can also be sour or bitter. It is produced by bees from the nectar of flowers, saccharine exudations from plants and also from the excretions of insects that suck plant sap; once collected, these substances are processed by the bees' own specific digestive enzymes and stored and matured in the combs of their hives to serve as food (BRASIL, 2000; ESCOBAR; XAVIER, 2013).

Honey is made up of hydrocarbons, sugars (mainly glucose and fructose), water, minerals, vitamins (C, D, E and B complex), enzymes, organic acids, aromatic substances, pigments, antioxidants (flavonoids and phenolics) and also pollen grain and traces of wax. Fungi, algae and solid particles from the stage of obtaining the raw material can also be found (BRASIL, 2000; MENDES et al., 2009; RICHTER et al., 2011).

Its organoleptic properties (aroma, color and taste) vary greatly depending on a number of factors, such as: diversity of vegetation, soil types, climatic conditions, bee species, handling and collection techniques of the beekeeper (SILVA et al., 2009).

It is a natural product that is highly acceptable on the market due to its nutritional and functional value. Historical records show that honey was not only part of the dietary habits of the first civilizations, but was also widespread in ethnomedicine. Currently, honey and its derivatives play an important role in therapeutic medicine, acting in the treatment of cancer, gastrointestinal disorders, asthma, cataracts, among other diseases (GOIS, 2011).

These characteristics are due to honey being a product of primarily botanical origin, with no additives or preservatives. As such, honey inherits many substances, particularly phenolic compounds, which authenticate it as an effective product in the treatment and prevention of many chronic-degenerative diseases.

With this in mind, the aim of this study was to evaluate the total content of phenolic compounds and the antioxidant capacity of honeys of the *Apis sp.* and *Melipona sp.* genera produced in the municipalities of Soure and Salvaterra - Marajó - Parà, in the belief that the results obtained in this study could support future research that legitimizes honey as a product with pharmacological, phytotherapeutic and functional properties.

1.1 The profile of beekeeping in Brazil

According to Camargo (2012), bees are considered social insects due to their order profile and specialization of work. They are extremely important for the balance and maintenance of the natural ecosystem, acting as pollinating agents for various plant species.

Of the products obtained from the hive, honey is considered the most important. Beekeeping meets the basic requirements of sustainability when it uses equipment and methods to better exploit the capacity of the bees, without causing them harm. It also combines economic benefits, as it generates income for producers, and ecological benefits, as it does away with the use of pesticides and deforestation (SOUZA, 2010).

In Brazil, beekeeping is divided into two very distinct practices: apiculture, which is the breeding of bees with sting, and meliponiculture, which is the management of "indigenous" bees without sting. The former, because it is more widespread, is the holder of the technology and production methods; meliponiculture, because it is still an artisanal branch, has no established standards for quality control and management to ensure that it is a product guaranteed to be marketed (CARVALHO et al., 2005).

Current Brazilian legislation, recommended by the Ministry of Agriculture and Supply (BRASIL, 2000), establishes quality standards only for honey produced by bees with stingers, partially meeting the characteristics of the product of stingless bees (STRAMM, 2011).

1.1.1 The beginnings of beekeeping in Brazil

Management of stinging bees began with the introduction of swarms from Spain and Portugal, which gave them the name "European bees". These insects had a docile temperament and adapted very well to the Brazilian climate. Around 1956, bees from East Africa arrived in Brazil, which in turn were very aggressive and caused many accidents during handling (PERALTA 2010).

Around the 1970s, beekeeping began to be practiced intensively in all the states, in line with the foreign market (PIRES, 2011). This progress led to the emergence of a hybrid, a species resulting from the crossing of African bees with European bees, resulting in an insect that inherited the resistance to pests of African bees and the large honey production of European bees (BERA, 2010).

1.1.2 Meliponiculture: a tradition of popular knowledge

Meliponiculture is the name given to the breeding of indigenous or native bees, popularly known as "stingless bees"; this predicate is due to the fact that these insects have atrophied stingers. Of those that occur in Brazil, the Melipona genus is the most widespread in the country's regions, and although other genera predominate in some places, the term "meliponiculture" has persisted to characterize this activity (NASCIMENTO, 2014).

Knowledge about indigenous stingless bees goes back a long way, as the native peoples of the Americas have been in contact with these insects for a long time. In colonial Brazil, at least until the 19th century, the honey of these bees was used by Indians and whites for the most diverse purposes and their wax for candle production. Descendants of the Mayan and Nahoa peoples in Mexico still maintain the customs of their ancestors, which consists of mixing honey with other substances to use as medicine (CARVALHO et al., 2005; VILLAS-BÔAS, 2012).

Meliponiculture is a practice that over time has not undergone technological advances, so that it is still developed in an artisanal way. It is a legacy of accumulated traditional knowledge that can be seen in the popular names of many species, such as: Jatai, Uruçu, Tiùba, Mombuca, Irapuà, Tataira, Jandaira, Guarupu, Manduri and many others (VILLAS-BÔAS, 2012).

According to Nascimento (2014), meliponineos produce honey in considerably smaller quantities than bees with stingers, but their honey is a target-consumer product, commanding the best prices on the market due to its particular aroma and flavor.

1.2 Honey and its botanical heritage

1.2.1 Obtaining the raw material and processing honey

Nectar is a secretion considered to be the raw material for honey production; it is obtained by bees from floral glands called nectaries. Its composition varies between species and is influenced by the age of the plant and soil and climate factors (SOUZA et al., 2012).

It is a solution containing sugars in an amount ranging from 5 to 80%, the rest being made up of amino acids, minerals, organic acids, vitamins and aromatic compounds; these substances give honey its basic properties (MOURA, 2010).

Honey processing begins with bees collecting nectar from flowers and secretions from other plants which, although they don't have flowers, secrete substances through the phloem that contain sugars, such as sugar cane. These insects use their mouth apparatus to collect it, passing through the pharynx, esophagus and stored in an organ anterior to the intestine, called the "bolsa melifera" (ESCOBAR; XAVIER, 2013; BERA, 2010).

Under these conditions, the chemical process is carried out by the action of enzymes (invertase, diastase, glucose oxidase, catalase and phosphatase) secreted by the bees themselves, which convert the complex sugars into glucose and fructose. When it reaches the hive, the result of this chemical process is deposited inside the combs, where the temperature is around 34 to 35° C (SILVA, 2006; STRAMM, 2011).

From then on, the physical process of honey maturation begins, which involves dehydration, concentrating the solid components. Finally, the process ends with the operculation (closing) of the combs, where the humidity is around 20% for bees with stingers (BERA, 2010).

1.2.2 The use of honey according to its bioactive properties

Despite the contributions of each species of bee, honey is primarily botanical in origin - as it is processed from the nectar of flowers and/or the saccharin secretions of other plants - bringing with it plant heritage. These include its bioactive properties due to the presence of phenolic compounds (PERALTA, 2010).

Bertoldi et al. (2012) suggest that the use of honey should not be limited to its sweetening properties, as it is also a functional food, meaning that in addition to its high nutritional value, it provides health benefits, such as the prevention of chronic degenerative diseases.

The relationship between a diet rich in antioxidant substances and the functional role they play in inhibiting the deleterious effects caused by free radicals has shown that these compounds have beneficial effects on health, such as inhibiting the appearance of cancer and cardiovascular diseases (PEREIRA, 2010).

These characteristics are due to the fact that honey contains numerous substances with antioxidant capacities, mainly phenolic compounds. In recent years, research into the identification of these compounds *in fresh* honey has increased considerably (OLIVEIRA et al., 2012).

In translations of Egyptian papyri, honey was part of the composition of hundreds of medicines used internally and externally to treat illnesses. In Babylon and Ancient Greece, the bodies of kings and generals killed in battle were bathed in honey to preserve them until the funeral (GÓIS, 2011).

Studies on the preservation of canine bones using honey have shown it to be effective, as it not only inhibits the action of pathogens, but also preserves bone rigidity (PIRES, 2011). Epithelial and corneal tissues preserved in honey had similar results, and the material is used for clinical purposes (BERTOLDI et al., 2012).

In remote regions, traditional medicine used honey to treat wounds topically, from cleansing to protection. It was also used to manage chronic wounds and treat airway disorders and cataracts. The indigenous peoples of Ceylon used honey to prolong the integrity of animal meat for up to a year (PERALTA, 2010).

1.3 The potential of antioxidants against the action of free radicals.

1.3.1 The harmful effects of free radicals

Free radicals are atoms or molecules that are unstable and highly reactive because they contain unpaired electrons. They can be formed as a result of natural metabolic processes in living organisms, such as cellular respiration (BARROS, 2011). They can also be formed to perform specific functions, such as fighting bacteria and viruses, producing energy, regulating cell growth and participating in the synthesis of many substances, among other functions (SA et al., 2012).

The problem arises when they are produced in excess, usually due to some biological dysfunction, causing damage to structural molecules in cells, which then becomes the initial mechanism for various diseases, such as cell death due to the rupture of their membranes, the inactivation of enzymes and a series of other harmful effects on the body (KIRSTEN; ZIMMERMANN, 2008).

In general, unpaired electrons are found mainly in oxygen and nitrogen atoms, both of which are strong oxidizers, making them the main free radicals responsible for oxidative stress. Radicals that come from reactive oxygen species are classified as "ROS", while those that come from reactive nitrogen species are called "RNS" (BARREIROS, 2006; SOUSA et al., 2007).

Among these main groups, hydroxyl ($^{\bullet}OH$) is considered to be the radical most harmful to the body, as it frequently attacks biomolecules by abstracting a hydrogen and/or adding it to the unsaturation. The attack on fatty acids, for example, occurs by subtracting a hydrogen from the methylene-allyl group, leading to the process of lipid peroxidation in cell membranes, which can result in a chain reaction, affecting countless other fatty acid molecules (BARREIROS; DAVID, 2006).

According to Oliveira et al. (2012), the oxidative damage suffered by these biomolecules is directly related to the causes of a large number of chronic diseases, such as premature ageing, cardiovascular diseases, cancer, immune system decline and neurodegenerative diseases, such as brain dysfunction. 1.3.2 The characteristics of an antioxidant substance

An antioxidant substance can be understood as one that has the ability to minimize, delay or inhibit oxidative reactions occurring in substrates. From a biological point of view, antioxidants can be understood as those substances capable of protecting biological systems against the damaging effects promoted by the action of free radicals on macromolecules and cellular structures (BARROS, 2011).

Antioxidants can basically be divided into two classes according to their action against oxidative damage. When these substances are capable of inhibiting or interrupting the chain of oxidative propagation in the substrate, they are classified as enzymatic antioxidants. The second class includes those capable of capturing radicals during the oxidation process and being consumed during the reaction, called non-enzymatic radicals (ANGELO; JORGE, 2007).

In terms of their origin, antioxidants can be of endogenous origin, which are those produced by the body itself, such as superoxide dismutase and peroxidase, which are enzymatic antioxidants, and oestrogenic sex hormones, which are non-enzymatic (SILVA et al., 2010).

Antioxidant species of exogenous origin are those that come from food, acting as "suicide" molecules; hence the need for continuous replenishment by eating foods that contain them. Among the best known are ascorbic acid (vitamin C), tocopherol (vitamin E), carotenoids and polyphenols (phenols and flavonoids) (ZIMMERMANN; KIRSTEN, 2008).

1.4 Phenolic compounds: secondary plant metabolites

1.4.1 General characteristics of phenolic compounds

Phenolic compounds are considered to be secondary plant metabolites, i.e. they are produced in response to environmental aggressions suffered by the plant or as a defense strategy against predators. More than 8,000 such substances have been catalogued in the plant kingdom and many are directly linked to the coloration of plants and fruits (SILVA et al., 2010).

The antioxidant properties of these structures are due to their ability to donate hydrogen or chelate metals, as well as the stability of the radicals formed (BARROS, 2011). Phenols are substances whose structure includes an aromatic ring with one or more hydroxyl substituents; figure 01 illustrates the basic structure of a phenol (ANGELO; JORGE, 2007).

Figure 01 - Basic structure of a phenol.

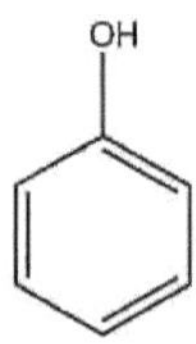

Source: authors (2017)

These compounds can be biosynthesized through different mechanisms, which is why they form a very distinct group from a metabolic point of view. In general, two metabolic pathways are considered to be involved in the synthesis of these compounds: shikimic acid and mevalonic acid. The former is the most involved in the biosynthesis of most plant phenols, which promotes the conversion of carbohydrates derived from glycolysis (ROSSATO, 2009).

These substances are very diverse and are divided into two large groups: flavonoids and non-flavonoids. The former is the group that deserves the most attention due to its high antioxidant activity (SILVA et al., 2010).

1.3.2 Flavonoids: chemical and structural behavior.

Flavonoids, also known as polyphenols, are substances that are not synthesized by the human body and are compounds peculiar to plants, especially in the aerial parts. In general, they are considered relatively resistant to climate change and oxidation processes (SILVA et al., 2015).

Polyphenols are made up of two aromatic rings which are interconnected by a central heterocyclic structure, the pyran. One of the benzenes is condensed with the sixth carbon of the pyran, while the other is in position 2 as a substituted phenyl group. Their carbons can undergo hydroxylation, hydrogenation, methylation and sulfonation reactions, making it possible to form more than four thousand different compounds (SILVA et al., 2015).

Figure 02 - Basic structure of a flavonoid

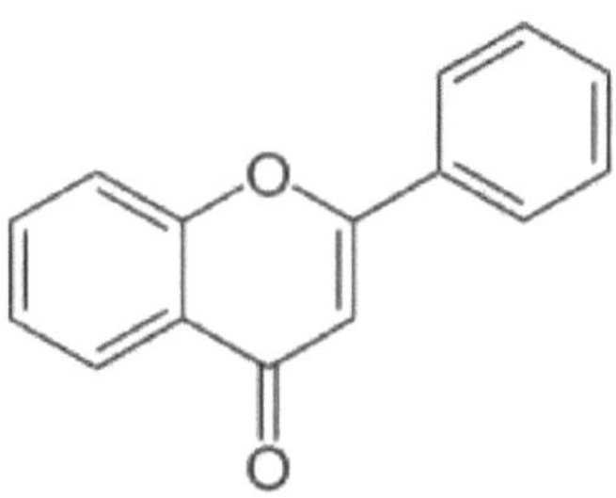

Source: authors (2017).

According to Zimmermann and Kirsten (2008), flavonoids represent a large group of polyphenols that are part of the composition of many fruits, vegetables and teas. They have been highlighted for their pharmacological properties. In this way, they are directly present in the human diet and act in the treatment and prevention of many chronic diseases.

Flavonoids have an ideal structure for capturing free radicals, and are considered even more effective than vitamin C and E. The antioxidant capacity of these compounds is due to their reactivity as hydrogen-donating agents, their ability to chelate transition metals and the stability of the flavanoyl radical formed. Generally speaking, the greater the number of hydroxyls, the greater the capacity to act as an electron-donating agent (ALVES et al., 2007).

Chapter 2

Research perspectives

To determine the total phenol content and antioxidant capacity of *Apis sp. and Melipona sp.* bee honeys produced in the municipalities of Soure and Salvaterra - Marajó - Parâ.

Check if there is a difference in the color of honey from bees with and without stingers;

© To quantify the content of total phenols in samples of honey from bees with and without sting using the Folin Ciocalteau method;

0 To evaluate the antioxidant potential of *Apis sp.* and *Melipona sp.* bee honeys using the free radical scavenging method (DPPH);

0 Compare the average results of the antioxidant capacity and total phenol content of *Apis sp.* and *Melipona sp.* honeys.

Chapter 3

Area of study

This study was carried out on 12 samples of honey *in natura,* with and without the presence of stingers, of the *Apis sp.* and *Melipona sp.* genera, respectively. The honeys came from the municipalities of Soure (latitude: 00° 43' 00" S; longitude: 48° 31' 24" W; altitude: 10m; Area: 3528.7 Km^2) and Salvaterra (latitude: 00° 45' 12" S; longitude: 48° 31'00" W; altitude: 5m; Area: 1048.2 Km^2), both located in the Arari micro-region, Marajó meso-region, State of Pará.

Figure 03 - Satellite image of the municipalities of Soure and Salvaterra - PA.

Source: Google Earth program version 7.1.5.1557 (2017).

The region's climate is basically hot and humid, with a tendency for rain to concentrate in the months from January to May, flooding the lower plains. In the following months (June to December) the region goes through a rigid period of drought, with exhausting summers

(BARBOSA, 2012).

3.1. Acquisition and identification of samples

The honey samples were purchased directly from producers in both municipalities. Part of each sample was transferred to hermetically sealed, sterile plastic containers in order to facilitate handling at the time of analysis and preserve the integrity of the rest of the samples. They were then taken to the Physico-Chemical Laboratory of the State University of Parà - Salvaterra Campus, where they were stored away from light and exhausting temperatures until the time of analysis. Figure 04 shows the samples in their original packaging.

Figure 04 - Twelve samples of honeys produced in the municipalities of Soure and Salvaterra, packed in their respective original packaging.

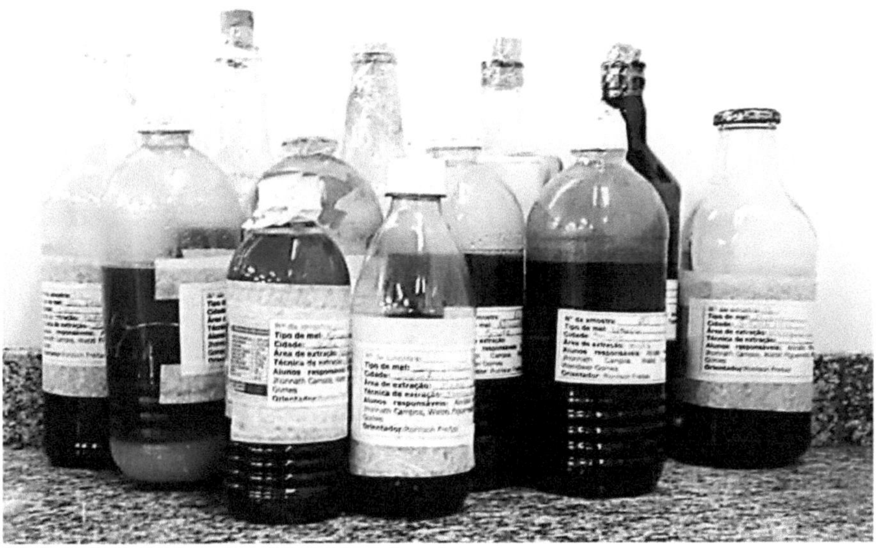

Source: authors (2017).

Table 01 (p. 28) characterizes each sample according to the information declared by the suppliers themselves: municipality of origin, extraction area and presence of iron.

Table 01 - Identification of honey samples according to municipality of origin, morphological characteristics of the harvest area, presence of iron and taxonomic genus.

Sample	Municipality	Collection area	Presence of iron	Gender
M1	Soure	Primary forest[1]	no	*Melipona sp.*
M2	Soure	Primary forest	no	*Melipona sp.*
M3	Salvaterra	Primary forest	no	*Melipona sp.*
M4	Soure	Primary forest	no	*Melipona sp.*
M5	Soure	Primary forest	no	*Melipona sp.*
M6	Salvaterra	Primary forest	no	*Melipona sp.*
A7	Soure	Primary forest	yes	*Apis sp.*
A8	Soure	Primary forest	yes	*Apis sp.*
A9	Salvaterra	Primary forest	yes	*Apis sp.*
A10	Soure	Mangrove[2]	yes	*Apis sp.*
A11	Salvaterra	Mangrove	yes	*Apis sp.*
A12	Soure	Primary forest	yes	*Apis sp.*

Source: authors (2017).

1 *Primary forest:* also known as virgin forest, this is an ecosystem that has never been significantly disturbed and/or altered by anthropogenic actions.
2 *Mangrove swamp:* a coastal ecosystem with a transition between terrestrial and marine environments.

Chapter 4

Experimental methods

4.1 Color determination

The colors of the honey samples were determined by adapting the method proposed by Biachi (1981). 5 g of honey were weighed and solubilized in 10 mL of distilled water, then the mixture was stirred in a vortex and filtered. The mixture was read in a UV-Vis spectrophotometer (Bel-photonics®) at 635 nm against a "blank" containing distilled water. The absorbance values (Abs_{635}) were substituted into the equation below and interpreted in terms of color according to the Pfund scale (Table 02).

$$mm = (371.39 \times Abs_{635}) - 38.70$$

Table 02 - Classification of honey color according to the Pfund scale.

Color	mm	Absorbance (635nm)
Water-white	0 -7,9	between 0.104 and 0.125
Extra white	8 - 16,4	between 0.125 and 0.148
White	16,5 - 33,9	between 0.148 and 0.195
Extra light amber	34 - 49,9	between 0.195 and 0.238
Light amber	50 - 84,9	between 0.238 and 0.333
Amber	85 - 113	between 0.333 and 0.411
Dark amber	114 or more	from 0.411

Source: Biachi (1981) adapted by the authors (2017).

4.2 Determination of Total Phenols

4.2.1 Principle of the Folin Ciocalteau method

Folin Ciocalteau's reagent is a solution of complexed ions formed from a heteropolyacid mixture of phosphomolybdic and phosphotungstic acid, where molybdenum is in the oxidized state (VI), responsible for the yellow color of the reagent. In the presence of reductants, the so-called blue molybdenum-tungsten complexes [$(PMoW_{I}\theta_{4})^{4-}$] are formed, with an average oxidation state between 5 and 6. Figure 05 (p. 30) illustrates the stage of the reaction which initially takes place with the deprotonation of the phenolic compounds promoted by the carbonate ion, forming phenolate ions. From then on there is an oxidation-reduction between the phenolate anion and the Folin Ciocalteau reagent, showing a blue color in the UV-Vis absorption range at 760 nm (OLIVEIRA et al., 2009).

Figure 05 - Molybdenum reduction reaction with gallic acid in a basic medium.

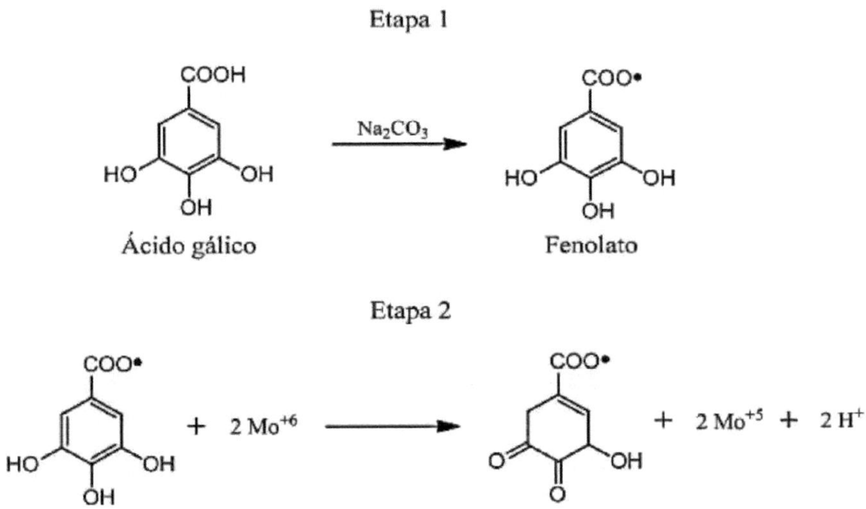

Source: Oliveira et al. (2009) adapted by the authors (2017).

4.2.2 Determination of total phenolics in the samples

To determine the total content of phenolic compounds, the method described by Wilczynska (2010) was adapted. Aqueous solutions of honeys were prepared at a concentration of 100 mg/mL. Aliquots of 750 μL of these solutions were mixed with 1.0 mL of 10% Folin-Ciocalteau reagent (Sigma-Aldrich®), after 3 minutes 1.0 mL of freshly prepared 7.5% sodium carbonate (Sigma-Aldrich®) and 10 mL of distilled water were added. The tests were carried out under cover of light. The mixture was homogenized in a vortex mixer and left to stand in the dark for 2 hours at room temperature.

After the reaction time, the mixture was read in a UV-Vis spectrophotometer at a wavelength of 760 nm against a "blank" containing all the reagents except the honey sample. The tests were carried out in triplicate for each sample analyzed, and the weighted average between them was taken. The results were expressed as equivalent milligrams of gallic acid per 100 grams of honey (mgEAG/100g honey). Figures 06 and 07 illustrate the change in color undergone by the Folin Ciocalteau reagent when reacting with the phenolic compounds present in the honey samples.

Figure 06 - Yellowish color of Folin Ciocalteau's reagent before reduction in the presence of honey solutions.

Source: Authors (2017).

Figure 07 - Bluish coloration of the Folin Ciocalteau reagent obtained after a two-hour reaction time with the honey samples.

Source: Authors (2017).

Ethanolic solutions of gallic acid (Sigma-Aldrich®) were prepared at concentrations ranging from 30 to 100 µg/mL. The absorbances obtained were plotted on a graph (x-axis) *versus* the respective concentrations of gallic acid (y-axis) using the Microsoft Office Excel 2013 program (Graph 01, p.33).

The curve generated was interpreted according to a linear trend line, obtaining the straight line equation ($y = 190.1x - 10.33$; $r^2 = 0.998$) which allowed us to indirectly determine the concentration of total phenolics in each 100 grams of sample through the absorbance presented at 760 nm. The data that generated the standard curve for gallic acid is shown below (Table 03).

Table 03 - Values of the different concentrations of gallic acid solutions, with their respective absorbances (triplicate), means and standard deviation.

Concentration	**Absorbance (760 nm)**			**-Mean ± SD**
	1ª analysis	**2nd analysis**	**3rd analysis**	
30 µg/mL	0,219	0,215	0,193	0,209 ± 0,01
40 µg/mL	0,278	0,261	0,283	0,274 ± 0,01
50 µg/mL	0,324	0,313	0,302	0,313 ± 0,01
60 µg/mL	0,381	0,381	0,362	0,375 ± 0,01
70 µg/mL	0,428	0,429	0,439	0,432 ± 0,00
80 µg/mL	0,478	0,485	0,481	0,481 ± 0,00

90 µg/mL	0,500	0,526	0,539	0,522 ± 0,02
100 µg/mL	0,585	0,567	0,583	0,580 ± 0,01

Source: Authors (2017).

Graph 01 - Standard curve for gallic acid.

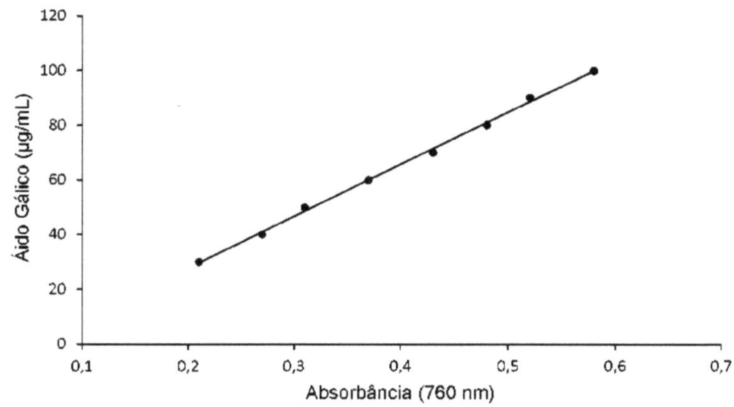

Source: Authors (2017).

4.2.4 Determination of antioxidant capacity using the DPPH method

42.4.1 Principle of the DPPH method

The DPPH radical (1,1-diphenyl-2-picrylhydrazyl), despite having an unpaired electron on the nitrogen atom, is a stable structure due to having three aromatic rings that promote the resonance effect, stabilizing its electronic charge. The purple color is due to the location of the free electron, which when reduced by an antioxidant becomes a yellow diamagnetic structure. The radicals formed by the antioxidant are not reactive enough to propagate a chain reaction; they are neutralized by reaction with another radical, forming stable products. The decrease in intensity of the purple color can be detected by the decrease in absorbance in the 515 - 520 nm range in the UV-Vis spectrum (SOUZA et al., 2015).

According to Oliveira (2015), the results obtained are generally presented as the amount of antioxidant needed to reduce the initial concentration of the radical by 50%. This value is achieved by graphing the percentage inhibition of DPPH as a function of the antioxidant concentration, which can result in a non-linear regression.

Figure 08 - Reaction between the DPPH radical and an antioxidant through hydrogen transfer.

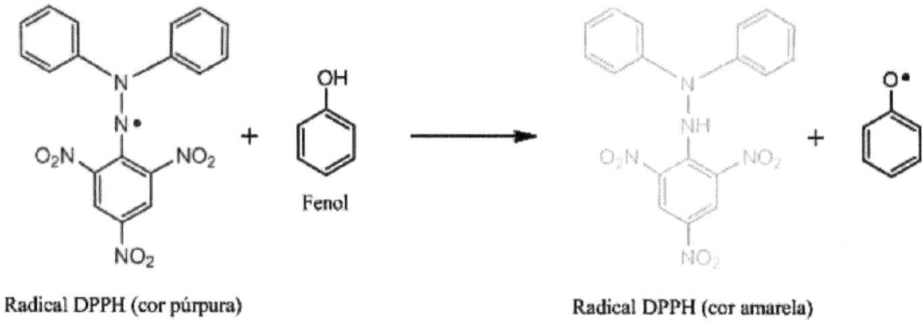

Source: Souza et al. (2015) adapted by the authors (2017).

4.2.4.2 Determination of antioxidant capacity in samples

To determine the antioxidant capacity of bee honeys, the method described by Pontis et al. (2014) was adapted. Initially, a methanolic solution of DPPH 100 µM was prepared and stored in an amber bottle, kept refrigerated at 4° C until the time of analysis. To carry out the tests, the solution was adjusted to an absorbance of around 0.625 with the addition of methanol.

Methanolic solutions of honeys ranging in concentration from 5 to 120 mg/mL were prepared. Aliquots of 2.1 mL of each concentration were mixed with 900 µL of DPPH. The mixture was homogenized and left to stand in the dark for thirty minutes. Together with the samples, a negative control was prepared, consisting of 900 µL of µM DPPH and 2.1 mL of methanol. After the reaction time, the samples and the negative control were read in a UV-Vis spectrophotometer at 517 nm against a "blank" containing only methanol. The absorbances were converted into percentage inhibition using the equation:

$$I_{(\%)} = [(Abs_{controle} - Abs_{amostra})/Abs_{controle}]x100$$

Where, $Abs_{control}$ is the absorbance of the negative control and Abs_{ample} is the absorbance of DPPH after reacting with the honey solution.

Using the Microsoft Office Excel 2013 program, the inhibition activity (y-axis) was plotted against the respective honey concentrations (x-axis). The curve generated was interpreted according to a polynomial trend line, obtaining the equation of the line ($y = ax^2 + bx + c$), which made it possible to estimate the concentration of honey effective in reducing the initial amount of DPPH by half (EC_{50}).

Figure 09 - DPPH radical staining after reaction time with different concentrations of honey (increase in concentration from left to right).

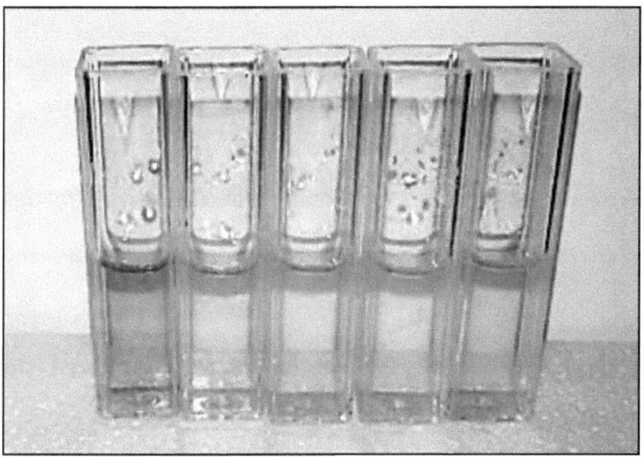

Source: Authors (2017).

4.2.4.3 Construction of the ascorbic acid standard curve

Methanolic solutions of ascorbic acid (Sigma-Aldrich®) ranging in concentration from 100 to 350 µg/mL were prepared. Aliquots of 20 µL of each concentration were mixed with 1980 µL of methanolic DPPH solution with an initial absorbance of approximately 0.625.

The mixtures were homogenized in a vortex and left to stand in the dark for thirty minutes. Together with the samples, a negative control was prepared containing 20 µL of methanol and 1980 µL of methanolic solution of DPPH radical. After the reaction time, the mixtures were read in a spectrophotometer

UV-Vis at 517 nm against a "blank" containing only methanol.

Figure 10 - Structural formula of ascorbic acid.

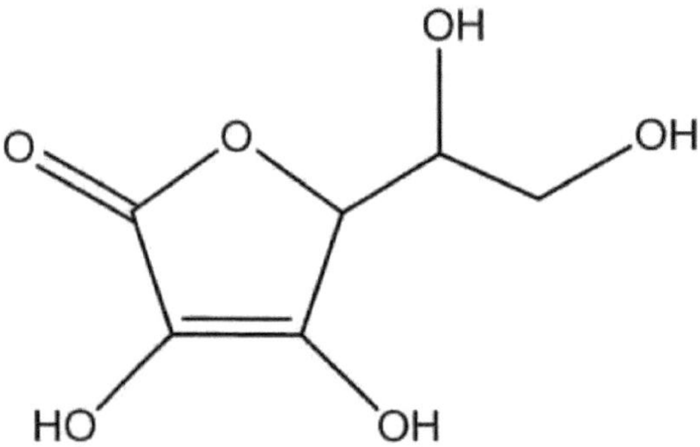

Source: Authors (2017).

In a graph, the inhibition activity (x-axis) was plotted *against* the respective concentrations of ascorbic acid (y-axis). The curve generated was interpreted according to a linear trend line, obtaining the straight line equation ($y = 3.396x + 17.47$; $r^2 = 0.990$). Table 04 shows the inhibition activity values found for each concentration considered, which allowed the standard ascorbic acid curve to be plotted (Graph 02, p. 37).

Table 04 - Values of the different concentrations of ascorbic acid solutions, with their respective inhibitory activity indices (triplicate), means and standard deviation.

Concentration	Inhibition Activity (%)			Mean ± SD
	1ª analysis	2nd analysis	3rd analysis	
100 µg/mL	24,76	24,44	24,28	24,49 ± 0,20
150 µg/mL	38,02	37,06	36,10	37,06 ± 0,78
250 µg/mL	55,59	55,91	55,91	55,80 ± 0,15

300 µg/mL	71,28	70,78	70,61	70,89 ± 0,29
350 µg/mL	80,57	80,91	79,90	80,46 ± 0,42

Source: Authors (2017).

Graph 02 - Standard curve for ascorbic acid.

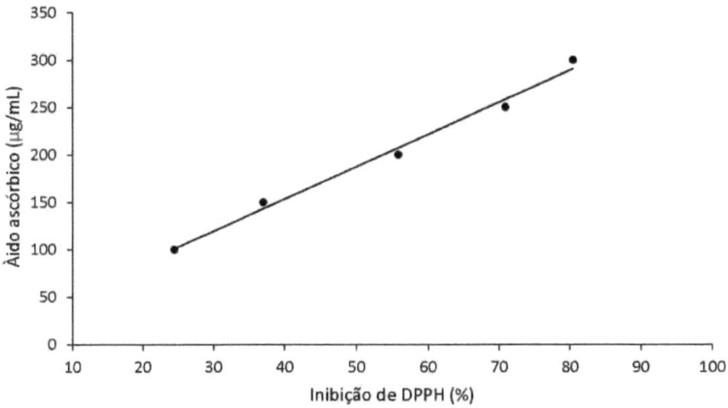

Source: Authors (2017).

Chapter 5

Results and discussion

The values (mean ± standard deviation) obtained for quantifying the total content of phenolic compounds and determining the color of the twelve samples of honey from bees with and without sting, produced in the municipalities of Soure and Salvaterra - Marajó - Parà, are shown below (Table 05):

Table 05 - Results of the color intensity and total phenols of the twelve samples analyzed.

Gender	**Samples**	**Color**	**mm**[1]	**Total Phenols**[2]
Melipona sp.	M1	dark amber	180,05 ± 0,80	20,59 ± 0,32
	M2	dark amber	125,95 ± 0,88	36,05 ± 0,56
	M3	dark amber	189,46 ± 0,97	51,71 ± 0,73
	M4	dark amber	235,76 ± 0,00	49,30 ± 0,50
	M5	dark amber	254,95 ± 0,46	51,52 ± 0,85
	M6	dark amber	160,37 ± 0,80	48,09 ± 0,32
Average	-	-	191,09	42,87
Minimum	-	-	160,37	20,59
Maximum	-	-	254,95	51,71
Gender	**Samples**	**Color**	**mm**[1]	**Total Phenols**[2]
Apis sp.	A7	amber	107,13 ± 0,76	36,05 ± 0,68
	A8	dark amber	188,96 ± 0,30	42,33 ± 0,41
	A9	dark amber	215,08 ± 0,88	50,76 ± 0,63
	A10	light amber	55,51 ± 0,35	31,87 ± 0,68
	A11	water-white	3,89 ± 0,35	48,85 ± 0,59
	A12	dark amber	145,76 ± 0,88	29,15 ± 0,32
Average	-	-	119,38	39,84
Minimum	-	-	3,89	29,15
Maximum	-	-	215,08	50,76

[1]color according to the millimeter scale proposed by Pfund (Biachi, 1981);[2] total phenol content expressed in milligrams equivalent of gallic acid per 100 grams of honey;

Source: Authors (2015).

5.1 Color

The color results found for the twelve samples of honeys analyzed show that there was a predominance of dark amber tones in 75% of the samples, followed by amber (8.33%), light amber (8.33%) and water-white (8.33%). When considering the six samples of *Melipona sp., it can* be seen that all of them were dark amber in color, which differs from the results found by Anacleto, Souza and Marchini (2009), who, when investigating samples of *T. angustula* honeys (stingless bees) produced in the municipality of Piracicaba, State of São Paulo, found colors ranging from extra light amber to amber.

Graph 03 - Percentage representation of the color shades found for the twelve samples of honey analyzed.

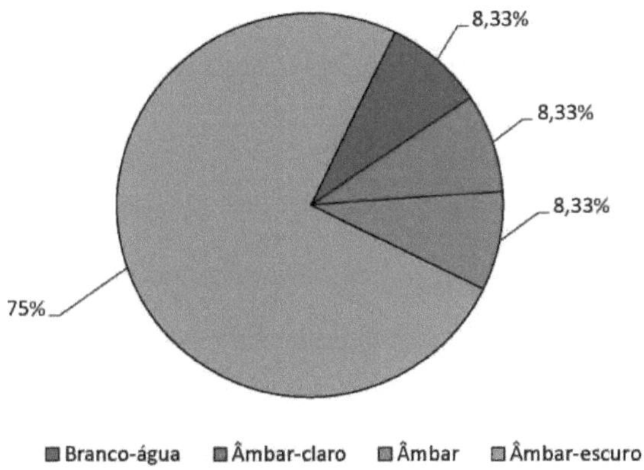

Source: Authors (2017).

For the honeys of *Apis sp.* bees, the findings are similar to those found by Barros et al. (2010), who, when evaluating the colors of honeys produced in the state of Rio de Janeiro, found variations between extra-light amber and amber, pointing out that these shades may vary from region to region, but that they are in accordance with Brazilian legislation.

In general, the results found in this study suggest that the color of the honeys does not necessarily depend on the type of bees, which could not yet be considered a determining factor for the honeys' hues. Silva et al. (2009) point out that the color of honey depends on the nectar and can range from almost colorless to very dark brown, depending only on the substances present in the plants the bees visit. According to Peralta (2010), the intensity of the color of honey is due to the pigments of phenolic compounds and he points out that darker honeys tend to have a higher content of these substances, which are related to the antioxidant properties of honey.

5.2 Total phenolic compounds

For the quantification of total phenols, values for *Melipona sp. ranged* from 20.59 to 51.71 mgEAG/100g of honey, with an average of 42.87. For *Apis sp.* honeys, values ranging from 29.15 to 50.76 milligram equivalents of gallic acid per 100 grams of honey were obtained, with an average value of 39.84. Lira et al. (2014), when quantifying the phenol content of *Apis mellifera*, found values ranging from 43.34 to 75.47 and for meliponineos (stingless bees) from 61.28 to 105.59 mgEAG/100g honey.

When comparing the amounts of phenolic compounds present in the honeys of bees with and without stingers analyzed in this study, it can be seen that they are slightly lower than the literature consulted. This may be linked to the climate and flora conditions, and substantially to the stress suffered by the plants providing the nectar.

This had already been proposed by Formiga et al. (2009), who hypothesized that the quantities of these compounds are directly linked to the stress conditions suffered by the plants that provide the raw material for honey.

According to Rocha et al. (2010), the varieties of phenolic compounds may also be linked to the large flow of nectar in a given vegetation, which prevents the bees from exploiting more sugar sources, resulting in honey with the contribution of only a few species.

There are no pre-established limits for the amounts of total phenols present in bee honeys, but the values found in current literature have an average range of less than or equal to 100 mgEAG per 100 grams of honey. This can be seen when we take as an example the results found by Pontis et al. (2014) who, when studying various samples of honeys produced in the state of Roraima, found concentrations ranging from 25 to 50.9 milligram equivalents of gallic acid per 100 grams of honey. In the west of Turkey, Ozkok, D'arcy and Sorkun (2010) found total phenolic contents ranging from 3.54 to 36.6 mgEAG/100g of honey.

When comparing the average content of phenolic compounds in bees with and without stingers, the *Box Plot* graph (Graph 04, p. 41) shows that there is a strong trend towards higher values in honeys from bees of the genus *Melipona sp.* This may indicate that although these insects exploit the same region, the species differ in terms of their radius of action and selection of nectar sources.

Graph 04 - Comparison of the average values of total phenols between the honeys of Apis sp. and Melipona sp. bees.

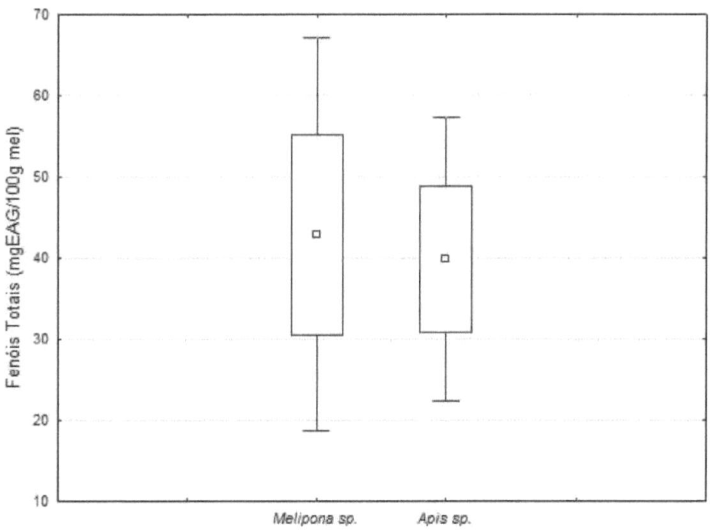

Source: Authors (2017).

Another point to consider is that the results for total phenols obtained in this study were even more significant than those found *in fresh* pulp by Faller and Fialho (2009), who, when determining the availability of phenolic compounds using the Folin Ciocalteau method in fruit and vegetables consumed in Brazil, found amounts of 15.3; 31.5; 45.1, 66.9 and 13.7 milligram equivalents of gallic acid per 100 grams of pulp.

These results show that honey does in fact inherit many of the properties of the plants from which it originated, and can form part of the dietary complement just as well as any other functional food, as it is a naturally organic product, free from additives and preservatives.

According to Silva (2015) phenols are to some extent linked to antioxidant capacity and suggests that the higher the concentration present in honey, the greater its anti-radical activity. However, for this to be predicted, a greater number of parameters need to be considered.

5.3 Evaluation of antioxidant capacity

Locatelli et al. (2009) in their work "*Study of the DPPH-scavenging activity: Development of a free software for the correct interpretation of data*" propose that it is necessary to evaluate the anti-radical capacity of a compound using *software*, as they believe that the usual trend lines may not efficiently satisfy the antioxidant behavior of some substances.

Some studies (ANDRADE et al., 2007; IHA et al., 2008; SA et al., 2012) estimate the minimum efficient concentration to capture 50% of the initial concentration of the DPPH radical using a linear regression line. However, this study suggests that, based on successive tests, the efficient concentration index can best be interpreted using a polynomial trend line.

This behavior had already been speculated by Oliveira (2015), who says that the DPPH method is widely used to determine the antioxidant activity of many biological compounds, which means that this method and its interpretation undergo intense modifications according to the characteristics of the drug being analyzed.

For example, graph 05 (p. 43) shows the anti-radical inhibitory potential of sample M4, where the points intersect along an exponential line, giving an $r^2 = 0.998$.

Graph 05 - Anti-radical behavior of sample M4 (Melipona sp.) interpreted according to a polynomial trend line.

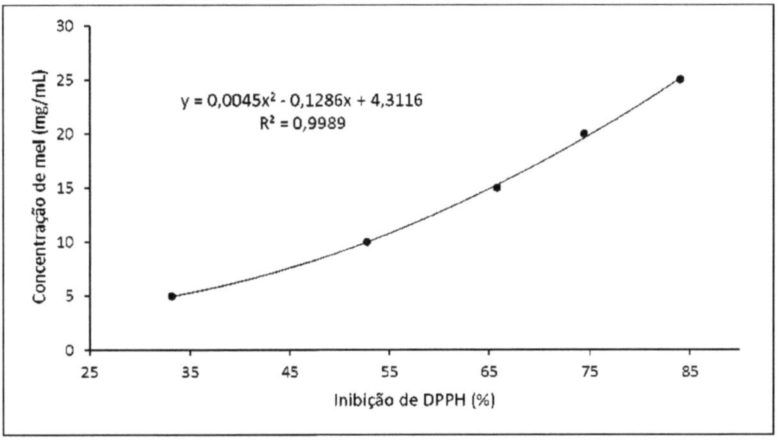

Fonte: Autores (2017).

The concentrations of honeys considered for the DPPH test and their respective percentage anti-radical inhibition indices are shown in tables 06 and 07 below. Also shown are the efficient concentration values, which were estimated according to the polynomial straight line equation obtained for each sample analyzed.

Table 06 - Concentrations, percentage inhibition index of the DPPH radical and efficient concentration of the *Melipona sp.* honey samples.

Samples	**Concentration (mg.mL)$^{-1}$**	**DPPH inhibition (%)**	CE50 **(mg.mL)$^{-1}$**
	40	26	
	60	36	
M1	80	41	109,71
	100	48	
	120	52	
	15	21	
	20	30	

M2	30	40	40,74
	40	50	
	60	60	
	5	10	
	10	17	
M3	20	23	101,26
	30	27	
	40	40	
	5	33	
	10	53	
M4	15	66	7,91
	20	74	
	25	84	
	5	18	
M5	10	37	
	15	48	15,89
	20	56	
	25	61	
	10	15	
	20	20	
M6	30	29	75,16
	40	38	
	50	45	

Source: Authors (2017).

When analyzing the data in the table above, we draw attention to samples M4 and M5, which showed the best DPPH free radical inhibition efficiency. Treatment with concentrations in the range of 5 to 25 mg/mL promoted an inhibition of 33 to 84% for the first sample and 18 to 61% for the second, with respective CE_{50} values of 7.91 and 15.89 mg/mL.

The tests with M1 and M3 showed the highest CE_{50} values, 109.71 and 101.26 mg/mL, respectively. This means that their antioxidant properties are inferior compared to the others. In the middle range of inhibition, M2 and M6 showed 40.74 and 75.16 mg/mL for the efficient inhibition concentration.

The results proposed in this work to assess the antioxidant capacity of *Melipona sp.* honeys were more satisfactory than the values found by Silva (2015), who, when checking the antioxidant potential of honeys produced *in* Rio Grande do Norte, obtained average CE50 values for *fresh* and dehumidified *Melipona subnitida* honey samples of 141.86 and 168.60 mg/mL, respectively.

Table 07 - Concentrations, percentage DPPH radical inhibition index and efficient concentration of *Apis sp.* honey samples.

Samples	Concentration (mg.mL)$^{-1}$	DPPH inhibition (%)	CE50 (mg.mL)$^{-1}$
	10	16	
	20	31	
A7	30	43	39,68
	40	48	
	50	59	
	5	9	
	10	13	
A8	20	20	72,79
	30	27	
	40	33	
	10	15	
	15	20	
A9	20	31	48,32
	25	44	
	30	54	
	10	15	
	20	21	
A10	25	25	78,54
	30	27	
	40	32	
	10	17	
A11	20	28	
	25	33	63,88
	30	36	
	40	40	

A12	10 20	13 24	43,11
	25	31	
	30	35	
	40	46	

Source: Authors (2017).

Table 07 shows the behavior of honey samples from stinging bees (*Apis sp.*) in terms of capturing the DPPH radical, where samples A7 and A12 showed the best antioxidant inhibition efficiency, with CE_{50} values of 39.68 and 43.11 mg/mL. Samples A9 and A11, on the other hand, showed intermediate levels of efficient concentration at 48.32 and 63.88 mg/mL, respectively. Lastly, A8 and A10 were similar in their ability to capture radicals, with respective CE_{50} values of 72.79 and 78.5 mg/mL, resulting in the lowest antioxidant activity of the group.

These results are similar to an investigation into the antioxidant capacity of *fresh Apis mellifera* honeys collected in Rio de Janeiro by Lianda (2009), who found CE_{50} values ranging from 10.81 to 52.64 mg/mL.

When comparing the results of the efficient concentration of the honey samples from *Apis sp.* and *Melipona sp.* bees, it can be seen that M4 and M5, belonging to the latter, were the ones with the highest anti-radical inhibition capacity. Mesquita et al. (2012), when working with honeys from *Apis mellifera* and *Melipona subnitida* bees, observed a higher antioxidant capacity in *Melipona* honey when compared to *Apis* honey.

However, when comparing the average CE_{50} of *Apis sp.* and *Melipona sp. honeys* using the *Box Plot* graph (Graph 06, p.47), it is clear that the honeys from bees with stingers still have the lowest concentration efficiency values, meaning that although two samples of *Melipona* sp. honeys stood out in terms of anti-radical action when compared to the others, *Apis sp.* honeys still have the highest antioxidant effectiveness.

This behavior has already been observed by Oliveira et al. (2012), who, when working with honeys from different locations in the Amazon, found an average of 22.3 mg/mL for *Apis mellifera* L. honeys and 31.5 mg/mL for *Melipona* species.

Graph 06 - Comparison of average EC50 values between *Apis* sp. and *Melipona* sp. bee honeys.

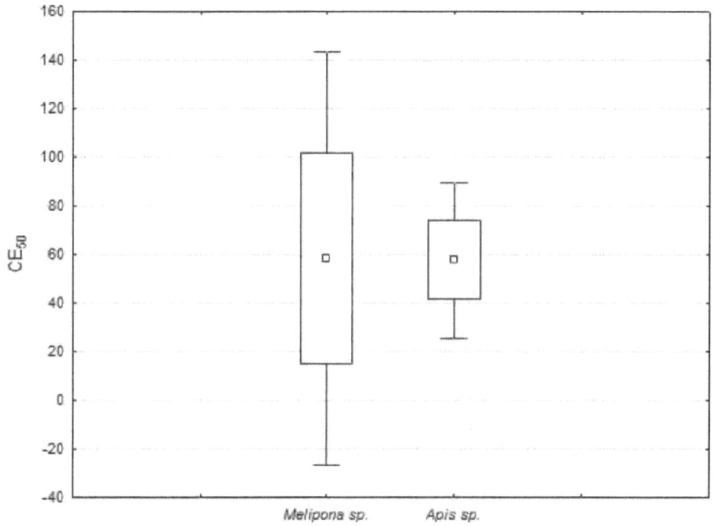

Source: Authors (2017).

We admit that there is a possible correlation between phenolic compounds and antioxidant activity, but this study needs to be further studied. This is because the data also shows samples that have considerable phenol values, but do not have satisfactory antioxidant activity, as was the case with samples M3 and A9. This does not detract from the validity of the results found in this study, but it does suggest a more in-depth study into the best conditions for capturing these radicals, bearing in mind that honeys have different conditions, such as pH range, for example.

It is also worth noting that, although some samples showed efficient antioxidant capacity, none matched the standard of ascorbic acid, which had a CE50 value of 0.1873 mg/mL. Pereira (2008) states that ascorbic acid, also called vitamin C, is a powerful antioxidant due to its ease of oxidation, as it contains a strongly reducing group in its structure, called reductone, which refers to the hydroxyls present in the C=C group.

Final considerations

The results obtained in this study show that although honey is a product that has been processed through several stages before reaching the consumer market, it still contains bioactive compounds that are inherited from the plants that provide the nectars. This enhances the product not only because of its nutritional value, but also because of its functionality.

Although only to a limited extent, the results obtained suggest the existence, to some extent, of a correlation between anti-free radical activity and the availability of phenolic compounds in honey, which has already been mentioned in the literature. However, there is a need for a more in-depth study of the parallel aspects that may act in favor of this manifestation.

The choice to evaluate the total phenol content and antioxidant capacity of honeys produced in the municipalities of Soure and Salvaterra is convenient both to establish a bioactive profile of the product sold in the region and to legitimize popular knowledge about the therapeutic properties of honey, which is still restricted to verbalized information without scientific proof.

We hope to provide subsidies for future research to be carried out in favor of a more specific characterization of this product and that its use as a functional food can be more widespread, given that it is a product linked to the prevention of many diseases caused by oxidative stress.

References

ANACLETO, D. A.; SOUZA, B. O.; MARCHINI, L. C. Composition of honey from the Jatai bee (*Tetragonisca angustula Latrille*, 1811). **Ciência e Tecnologia de Alimentos**, v. 29, n. 3, p. 535-541, 2009.

ANDRADE, C. A.; COSTA, C. K.; BORA, K.; MIGUEL, M. D.; MIGUEL, O. G.; KERBER, V. A. Determination of phenolic content and evaluation of antioxidant activity of Acacia Podalyriifolia A. Cunn ex. G. Don, Leguminosae-mimosoideae. **Brazilian Journal of Pharmacognosy**, v. 17, n. 2, p. 231-235, 2007.

BARBOSA, M. J. S. (Org.). **Analytical report of the Marajó territory**. Belém: MDA/ PITCPES-UFPA/GPTDA, 2012. Available at: <http://sit.mda.gov.br/download/ra/ra129.pdf>. Accessed on: December 12, 2015.

BARREIROS, A. L. B. S.; DAVID, J. M. Oxidative stress: relationship between generation of reactive species and body defense. **Revista Quimica Nova**, v. 29, n. 1, p. 113123, 2006.

BARROS, L. B.; TORRES F. R.; AZEVEDO, L. C.; BART, O. M.; FREITAS M. Q. Physico-chemical characterization of honey produced by *Apis mellifera* in the state of Rio de Janeiro. **Revista Brasileira de Ciência Veterinària**, v. 17, n. 34, p. 117-120, 2010.

BARROS, Z. M. P. **Tropical fruit peels as a source of antioxidants for enriching ready-made juice**. 2011. 85 f. Dissertation (Master's Degree in Food Science and Technology) - University of São Paulo, Piracicaba, 2010.

BERA, A. **Effects on physico-chemical, microbiological and sensory characteristics of honey samples from bees subjected to gamma radiation**. 2010. 112 f. Thesis (Doctorate in Nuclear Technology: Applications) - University of Sao Paulo, Sao Paulo, 2010.

BERTOLDI, F. C.; GONZAGA, L. V.; FETT, R.; REIS, V. D. A. Evaluation of antioxidant activity and determination of total phenolic compounds of honeys produced in the Pantanal. **Revista Evidência**, v. 12, n. 2, p. 155-164, 2012.

BIACHI, E. M. **La miel, caracteristicas y composición:** Anàlisis y Adulteraciones. Santiago del Estero: UNSE - CEDIA, 1981.

BRAZIL. Ministry of Agriculture, Livestock and Supply. **Normative Instruction No. 11, of October 20, 2000.** Establishes the technical regulation for the identity and quality of honey. Official Gazette of the Federative Republic of Brazil, Brasilia, October 23, 2000. Section 1, p.16 - 17.

CAMARGO, I. M. **Healing potential of bee honey (*Apis mellifera* L.) in skin tissue lesions of wistar rats**. 2012. 47 f. Monograph (Pharmacy Course) - Assis Gurgacz College, Cascavel, 2012.

CARVALHO, C. A.; SOUZA, B. A.; SODRÉ, G. S.; MARCHINI, L. C.; ALVES, R. M. O. **Honey from stingless bees:** a contribution to physico-chemical characterization. 1 ed. Cruz das Almas: Nova Civilizaçâo, 2005.

ESCOBAR, A. L. S.; XAVIER, F. B. Physiotherapeutic properties of bee honey. **Revista Uningà**, n. 37, p. 159-172, 2013.

FALLER, A. L. K.; FIALHO, E. Availability of polyphenols in fruits and vegetables consumed in Brazil. **Revista Saùde Pùblica**, v. 43, n. 2, p. 211-218, 2009.

FORMIGA, A. T.; GONÇALVES, S. J. M. R.; SOARES, G. L. G.; ISAIAS, R. M. S.; Relationships between total phenol content and the cycle of *Cecidomyiidae* galls on *Aspidosperma spruceanum Mull. Arg*. (Aponinaceae). **Acta Botànica Brasilica**, v. 23, n. 1, p. 93-99, 2009.

GOIS, G. C. **Physico-chemical characterization and microbiological quality of *Apis***

***melifera* honey sold in the state of Paraiba**. 2011. 102 f. Dissertation (Master's Degree in Animal Science) - Federal University of Paraiba, Areia, 2011.

GÓIS, G. C.; LIMA, C. A. B.; SILVA, L. T.; EVANGELISTA-RODRIGUES, A. *Apis mellifera* honey compound: quality requirements. **Acta Veterinaria Brasilica,** v. 7, n. 2, p. 137-147, 2013.

IHA, S. M.; MIGLIATO, K. F.; VELLOSA, J. C. R.; SACRAMENTO, L. V. S.; R.; PIETRO, C. L. R.; ISAAC, V. L. B. I.; BRUNETTI, I. L. B.; CORRÊA, M. A. C.SALGADO, H. R. N. S. Phytochemical study of guava (*Psidiumguajava L.*) with antioxidant potential for the development of phytocosmetic formulations. **Brazilian Journal of Phamacognosy**, v. 18, n. 3, p. 387-393, 2008.

LIANDA, R. L. P. **Profile of phenolic substances in Brazilian honeys by high performance liquid chromatography and evaluation of their antioxidant potential**. 2009. 156 f. Thesis (Doctorate) - Federal Rural University of Rio de Janeiro, Seropédica, 2009.

LIRA, A. F. SOUSA, J. P. L. M; LORENZON, M. C. A.; VIANNA, C. A. F. J.; CASTRO, R. N. Comparative study of *Apis mellifera* honey with Meliponineos honeys. **Acta Veterinaria Brasilica**, v. 8, n. 3, p. 169-178, 2014.

LOCATELLI, M.; GINDRO, R.; TRAVAGLIA, F.; COISSON, J.; RINALDI, M.; ARLORIO, M. Study of the DPPH-scavenging activity: Development of a free software for the correct interpretation of data. **Food Chemistry,** v. 114, n. 3, p. 889-897, 2009.

MENDES, C. G.; SILVA, J. B. A. MESQUITA, L. X.; MARACAJA, P. B. Honey analysis: a review. **Revista Caatinga**, v. 22, n. 2, p. 07-14, 2009.

MOURA, S. G. **Good beekeeping practices and the quality of honey from *Apis melifera***

***Linnaeus*, 1758**. 2010. 94 f. Thesis (Doctorate in Animal Health and Reproduction) - Federal University of Piaui, Teresina, 2010.

NASCIMENTO, A. S. **Physico-chemical parameters, pollination and trace element determination of *Meliponinae* honey (Hymenoptera: Apidae)**. 2014. 114 f. Thesis (Doctorate in Entomology) - University of Sao Paulo, Piracicaba, 2014.

OLIVEIRA, A. C.; VALENTIM, I. B.; GOULART, M. O. F.; SILVA, C. A.; BECHARA, E. J. H.; TREVISAN, M. T. S. Natural plant sources of antioxidants. **Quimica Nova**, v. 32, n. 3, p. 689-702, 2009.

OLIVEIRA, P. S.; MÜLLER, R. C. S.; DANTAS, K. G. F.; ALVES, C. N. Phenolic acids, flavonoids and antioxidant activity in honeys *of Melipona fasciculata, M. flavolineta* (*Apidae*, *Meliponini*) and *Apis mellifera* (*Apidae*, *Apini*) from Amazonia. **Quimica Nova**, v. 35, n. 9, p. 1728-1732, 2012.

OZKOK, A.; D'ARCY, B.; SORKUN, K. Total Phenolic Acid and Total Flavonoid Content of Turkish Pine Honeydew Honey. **Journal of Product and Medical Science**, v. 2, n. 2, p. 65-71, 2010.

PERALTA, E. D. **Antimicrobial activity and chemical composition of honeys from the State of Bahia**. 2010. 267 f. Thesis (Doctorate in Biotechnology) - State University of Feira de Santana, Feira de Santana, 2010.

PEREIRA, M. A. **Chromatographic profile of phenolic substances present in extracts of fish roast honey and evaluation of their antioxidant power**. 2010. 77 f. Monograph (Degree in Chemistry) - Federal Rural University of Rio de Janeiro, Seropédica, 2010.

PEREIRA, V. R. **Ascorbic acid:** characteristics, mechanisms of action and applications in the food industry. 2008. 40 f. Monograph (Bachelor's Degree in Food Chemistry) - Federal

University of Pelotas, Pelotas, 2008.

PIRES, R. M. C. **Quality of honey from *Apis melifera* Linnaeus, 1758 bees produced in Piaui**. 2011. 94 f. Dissertation (Master's Degree in Food and Nutrition) - Federal University of Piaui, Teresina, 2008.

PONTIS, J. A.; COSTA, L. A. M. A.; SILVA, S. J. R.; FLACH, A. Color, phenolic and flavonoid content, and antioxidant activity of honey from Roraima, Brazil. ***Food Science and Technology***, Campinas, v. 34, p. 69-73, 2014.

RICHTER, W.; JANSEN, C.; VENZKE, T. S. L.; MENDONÇA, C. R. B.; BORGES, C. D. Evaluation of the physicochemical quality of honey produced in the city of Pelotas/RS. **Alimentos e Nutriçao**, v. 22, n. 4, p. 547-553, 2011.

ROCHA, H. C.; LARA, A. A.; CECCHETTI, D.; PACHECO, A. B. Physico-chemical characteristics of honeys produced in combs of different ages. **Ciências Agràrias**, v. 31, n. 3, p. 639-644, 2010.

ROSSATO, S B. **Antioxidant potential and phenolic compounds of peaches (*Plunus persica L. Batsch*).** 2009. 61 f. Thesis (Doctorate in Pharmaceutical Sciences) - Federal University of Rio Grande do Sul, Porto Alegre, 2009.

SA, P. G. S.; GUIMARAES, A. L.; OLIVEIRA, A. P.; FILHO, J. A. S.; FONTANA, A. P.; DAMASCENO, P. K. F.; BRANCO, C. R. C.; BRANCO, A.; ALMEIDA, J. R. G. S. Total phenols and antioxidant activity of *Selaginella convoluta (Arn.) Spring* (*Selaginellaceae*). **Revista de Ciências Farmacêuticas Bàsica e Aplicada**, v. 33, n.4, p. 561-566, 2012.

SILVA, E. V. C. **Characterization and pasteurization of honeys from *Melipona fasciculata* (grey annatto) and *Apis mellifera* (Africanized) bees**. 2006. 51 f.

Dissertation (Master's Degree in Food Science and Technology) - Federal University of

Pará, Belém, 2006.

SILVA, M. C. P. **Physico-chemical characterization, antioxidant content and sensory profile of bee honeys submitted to dehumidification and humidification**. 2015. 81 f. Dissertation (Master's in Animal Production) - Universidade Rural do Semi-Arido, Mossoró, 2015.

SILVA, R. A.; AQUINO, I. S.; RODRIGUES, A. E.; SOUZA, D. L. Physico-chemical analysis of honey samples from zamboque bees (Frieseomelita Varia) from the Seridó region of Rio Grande do Norte. **Revista Verde**, v. 4, n. 4, p. 70-75, 2009.

SOUSA, C. M. M.; SILVA, H. R.; VIEIRA JÛNIOR, G. M.; AYRES, M. C. C.; DA COSTA, C. L. S.; ARAÙJO, D. S.; CAVALCANTE, L. C. D.; BARROS, E. D. S.; ARAÙJO, P. B. M.; BRANDAO, M. S.; CHAVES, M. H. Total phenols and antioxidant activity of five medicinal plants. **Revista Quimica Nova**, vol. 30. n. 2, p. 351355, 2007.

SOUZA, E. H.; MASSARIOLI, A.; MORENO, I. A. M.; ALENCAR, S. M.; SOUZA, F. V. D.; MARTINELLI, A. P. Quantification and composition of nectar in *Bromeliceae*. In: BRAZILIAN CONGRESS ON GENETIC RESOURCES, 2., Belém, 2012. **Proceedings...**Belém, p. 24-26, 2012.

SOUZA, R. F. **Physico-chemical, biochemical and microbiological parameters as descriptors of quality and discrimination of honeys from the State of Parâ**. 2010. 121 f. Dissertation (Master's in Chemistry) - Federal University of Parâ, Belém, 2010.

STRAM, K. M. **Composition and quality of Jandaira bee honeys (Melipona subnitida), storage effects and comparison with *Apis mellifera* honeys**. 2011. 106 f. Dissertation (Master's in Pharmaceutical Sciences) - University of São Paulo, São Paulo, 2011.

WILCZYNSKA, A. Phenolic content and antioxidant activity of different types of polish honey

- a short report. **Polish Journal of Food and Nutrition Sciences**, v. 60, n. 4, p. 309-313, 2010.

ZIMMERMANN, A. M.; KIRSTEN, V. R. Foods with antioxidant function in chronic diseases: a clinical approach. **Disciplinarum scientia**, Santa Maria, v. 9, n.1, p. 51-68, 2008.

MIX
Papier aus verantwortungsvollen Quellen
Paper from responsible sources
FSC® C105338

Printed by Books on Demand GmbH, Norderstedt / Germany